William Alazawi
&
Laura Maria Vearncombe

Ready to Excel: Secrets for the New Doctor

While every effort is made to ensure that the information given in this text is accurate, medical knowledge is constantly changing and no responsibility for loss or damage howsoever occasioned to any person acting or refraining from action as a result of information contained herein is accepted by the authors or publishers. Readers are strongly advised to confirm that the information complies with the latest legislation and standards of practice.

ISBN 978-1-4457-9177-7

Foreword

The first month of the first Foundation year probably involves the most rapid learning and the biggest change of life that a doctor will encounter. Not surprisingly, this period will usually be a mixture of excitement and stress, hopefully leading to an increasing sense of achievement and that the right career has actually been chosen. Even though medical school curricula have changed and developed hugely over the years, the transition to being a practising doctor with real responsibility will always have the potential to be difficult and stressful, and indeed if it were not so, there might be a tendency to undervalue the importance of the job, and the huge trust that the public puts in the medical profession.

This book, edited by two doctors who have recently gone through the transition, will give invaluable practical guidance to those about to go through the same experience. It highlights all the essentials of the Foundation years, including changing from a student to doctor, helping to give patients the best possible care, minimising stress on oneself and others, learning while working, and planning a career. It can be strongly recommended to final year medical students who are building up to their Foundation Years, and the Foundation doctors themselves.

Dr Michael Glynn MA MD FRCP FHEA
Consultant Physician and
Gastroenterologist/Hepatologist
Barts and the London NHS Trust
Director - North East Thames Foundation School

Contributing Authors

Dr William Alazawi MA(Cantab) MB BChir PhD MRCP
Academic Specialist Registrar in Gastroenterology & Hepatology
Barts & The London School of Medicine

Dr Laura Maria Vearncombe BMedSci MBBS MRCP
General Medicine Specialty Registrar
Wessex Deanery

Dr Natalie Direkze MA(Cantab) MBBS MRCP PhD
Consultant Physician & Gastroenterologist
Frimley Park Hospital

Dr Sarah George MBBS
Radiology Specialty Registrar
London Deanery

Dr Adam Graham MBBS
Foundation Doctor
North East Thames Foundation School

Dr Farzana Hakim MA(Cantab) MB BChir
Paediatrics Specialty Trainee
Severn Deanery

Dr Ashnila Janmohammed MBBS
Foundation Doctor
North East Thames Foundation School

Dr Farrah Jawad BSc MBBS
Foundation Doctor
North East Thames Foundation School

Dr Noor Jawad BSc MBBS MRCP
Specialist Registrar in Gastroenterology
Clinical Research Fellow
Cancer Research UK

Mr Amit Shah MA(Cantab) MB BChir MRCS
Specialist Registrar in Plastic Surgery
Eastern Deanery

Contents

Introduction

As a new junior doctor, fresh out of medical school there will be a lot of firsts. First take, first pre-clerking clinic, first ward cover, first crash call, first post-take ward round to run…

In the blur that is your first year as a doctor, it can be difficult to appreciate your role in the complex organisation of the hospital. It takes a little while to catch on to seemingly simple things that can make your life easier; building the right working relationships or knowing where to go for help.

However, the most important components of your learning are how to be safe and efficient in managing your patients' hospital stay on a day to day basis as the doctor 'at the coal-face' and how to treat those sick patients you come across in emergency situations.

Take pride in your work. Understand that you are on the bottom rung of the apprenticeship ladder, and that some of the tasks assigned to you may indeed seem basic. Nevertheless, they are vital to patient safety and are the firm foundations for your development.

On occasion, your responsibilities to your job and to your training can start to feel overwhelming, but within only a few months of starting you will look back and realise what a steep learning curve you've come through.

The purpose of this guide is not to re-hash how to take a history and examine patients – you should be well equipped to do that already. Medical school should have provided you with the basics, and there is nothing like experience to make you slick at your job. However, in this book we will pass on to you the tips we have learned so that you hit the ground running.

Routinely, your role will be to manage the progress of the investigation and treatment of the patients admitted under your Consultant's care. You will be the first port of call for the patient and their family and for the ward staff when there are problems.

However, being a good House Officer is not simply about getting jobs done. You will want to grow into a capable and trusted doctor. This means the ability to:

Take responsibility where appropriate.

Be safe.

Be thorough.

Be clear about difficult issues.

Assume nothing.

Ensure a 'safety net' for your decisions.

Be alert to non-clinical problems that impact on your patients.

We will show you how to integrate these principles into your daily work and how they are the basis of the 'tricks of the trade' that each generation of junior doctor has had to cobble together for itself along the way. For many, the learning curve is dominated by episodes of poor communication, snappy Registrars annoyed at badly made referrals, meteoric stress levels when on call, and worst of all, delays and compromise to patient care.

Whilst there is no shortcut to experience, why not start your career with these principles in your grasp instead of learning the hard way? Use this guide to get into good habits right from the beginning so that your work pattern is thorough and safe throughout your career. Be as effective and as efficient as possible right from the word "Go".

This book will show you how.

1. Making Ward Life Work

As a general medical or surgical House Officer, the ward is where the bulk of your day-to-day professional responsibilities lie and where the bulk of your learning (and career progression) opportunities lie also.

First impressions really do count. Even more than making sure you dress appropriately the single most important thing for you to do is to get off on the right foot with your colleagues. Most House Officers now start with a short shadowing period. If your rotation does not offer one as a matter of course – ask for one. Commitment to this induction period is an essential part of demonstrating to your team that you are diligent, to the staff on your ward that you are approachable and to yourself that you will be ready to treat your patients on Day One.

Use this time to familiarize yourself with your physical surroundings (the crash trolley, the notes trolley, where cannulae are kept and so forth). Get to know your new counterpart; the other House Officer on the firm. A good working relationship with them will make a real difference to your experiences of the post and the importance of the support you will give each other is immeasurable.

Introduce yourself to the ward staff, in addition to your own medical firm. The ward staff will include the Charge Nurse or Sister (the Ward Manager), the nurses and the ward clerk, as well as physiotherapists and occupational therapists, all vital members of the ward's Multi-Disciplinary Team (MDT). This is when the Ward Manager will tell you what they expect of you and advise you of some of your responsibilities. It is worthwhile to take heed and to nurture a good professional relationship with nurses and other non-medical staff. Having the nursing staff on your side when you are particularly busy - while doing a ward cover on-call, for example - will be invaluable (see Chapter 5 Being On-Call). They will be able to prioritise the jobs you are needed for and may even help you with some practical procedures.

The start of your working day will usually be the ward round, for which you will have prepared an adequately detailed list of your patients (see Chapter 4 The Ward Round). Having accustomed yourself to your pattern of working life, the ward round will hopefully be an opportunity to shine. Often, due to the nature of on-call rotas, you may be the only member of the team who knows the patients, and the latest updates in their management plans. You will therefore have the opportunity here, not only to display your organisational skills in ensuring a smoothly run ward round, but you can also stand out by demonstrating excellence in your clinical presentation skills. These should be practised and perfected on a daily basis whether with your own firm or when referring to other specialist teams. Learning to perform well in this forum will always stand you in good stead for the future. Your enthusiasm and efficiency will be noted and in turn you will reap the benefits, with your seniors supporting and encouraging your professional development.

After the ward round, you should get together with your team to run through the jobs that need doing for each patient, and then share out your list of tasks appropriately between yourselves. It may well be worth arranging a time to reconvene well before the end of the working day to assess what still remains to be done, and to help each other out accordingly. Any problems that have arisen can also be addressed at this point. At the end of the day, you can meet again to round-up, update your patient list ready for the following day and complete routine tasks such as requesting phlebotomy for the following morning. The on-call doctor will need to receive a proper verbal and written handover so that any jobs for the evening or night can be addressed, and so that any particularly unwell patients are brought to their attention. Try to ensure that these informal meetings, where your patients are discussed, take place in a private and undisturbed environment such as the doctors' office or another appropriate venue.

In addition to your daily responsibilities of the ward round and its resulting tasks for each patient, there will be other activities, often running on a weekly basis, which you will need to consider when planning out each day. Almost certainly, you will be expected to attend the weekly Ward Multidisciplinary Team meeting and provide a medical summary or update for each of your patients that are discussed. In addition, the Sister in charge, physiotherapists, occupational therapists and social workers who have been working with your patients will contribute. This meeting provides the basis for discharge planning; a process that should begin from the admission date of each patient, and is particularly relevant among elderly patients, or those whose social circumstances change as a result of their admission to hospital. It may well be your responsibility to create the list of patients who need to be discussed at this meeting, and submit it in advance of the session. You may also need to provide a summary of the outcomes of this meeting, or at least be able to communicate such information back to your team.

Unless there is an emergency or an unexpected situation amongst your team members, you should try to manage your time efficiently so that you can attend meetings and, in particular, designated teaching sessions punctually. These training commitments will need to be borne in mind when dividing up the list of jobs with your colleagues. For example, you will need to be aware of when there is scheduled teaching for yourself, and also for your (possibly senior) colleagues so that you can ensure that the appropriate cover is available to deal with the tasks of the day. Whenever possible, it is worth your while attending other educational or clinical meetings, for example Morning Report, Grand Round, cancer multidisciplinary meetings and joint meetings with the Radiology or Pathology departments. These will broaden your educational experience, perhaps inspire interest in other specialties and you may also have the opportunity to present your own patients in a more formal setting with a bigger audience.

The other major responsibility you are likely to have, at some point during your rotation, is teaching and guiding medical students. Final year students value time spent with the House Officer in particular, as they will be expected to perform the same job very shortly, and they too need to learn how to deal with life on the wards. There will be scheduled teaching sessions already incorporated into their own timetables, but as well as involving them in carrying out the ward jobs, it is a good idea to provide some formal teaching yourself. They will appreciate your input as the doctor who has most recently gone through the experience of Finals. Getting involved in teaching is a valuable and very rewarding experience, and you may wish to take this role further as your career progresses (see Chapter 8 What Can House Officers Do To Progress In Their Career?).

The more you get involved in these activities and take up other opportunities such as audit and research within your department, the better equipped you are for achieving success further along in your career. You should have the chance to give at least one formal presentation during each rotation, whether a case study or even an audit or research project you have been involved in. If this is not the case, then this issue ought to be discussed with your Educational Supervisor,

Head of Department or the Foundation Year co-ordinator within your Trust. In as much as you are responsible for your own professional development, your seniors have a role in providing support to help you achieve your personal objectives. If you are concerned about this, you should highlight the situation promptly.

The nursing staff on the ward will usually be happy to help when it comes to learning basic practical skills that they can teach you. In areas that you feel you lack experience (for example naso-gastric tube insertion), you may ask the nurses to bleep you when such an opportunity arises. Avoid leaving completion of your competency assessments until the end of your placement, when you may find that there are many training doctors rushing around trying to find seniors to get "signed-off". If your Deanery uses paper rather than on-line assessment forms, have a few copies of these to hand at all times (perhaps in your folder) so as to complete them as you go along. Additionally, keep a note of interesting cases on a regular basis so that you can consider using them for presentations, or assessments at a later stage. Ask your seniors for opportunities to observe and then perform more complex practical procedures; a request they are usually more than happy to facilitate. As well as increasing your own skill base, this will also help to make your portfolio stand out from those that are simply filled with "standard" House Officer procedures.

The ward provides a wealth of excellent basic experience for you, the newly-qualified doctor, given the range of patients you will be admitting and the various healthcare professionals working alongside you. Unsurprisingly, the ward can be a stressful environment for the busy junior doctor, and there will be moments where you may feel you practically live there! The key to making the most of these stressful periods is in seeking support when you need it, and making sure that the workload is fairly distributed. Remember also that you are not tied to the ward - you are allowed to leave it and venture into the wider hospital out there!

2. The Personal Side of the House Officer Year

The transition from medical student to House Officer is a big change. A new doctor will invariably undergo something of a lifestyle change in getting to grips with the new job. Finding time for things in your personal life becomes trickier. Opportunities you possibly took for granted as a student are no longer available, such as being able to pop to the bank or post office between lectures, getting the engineer to come round to fix the washing machine between 9-5 on a weekday, and so on and so forth. You even had time to do things you enjoy like play sport, music, or participate in student societies. Suddenly, when you start work, finding the time to do all of these things becomes a greater challenge.

Don't feel discouraged! It is all about being organised. The chances are that if you got through medical school relatively unscathed, you have the skills to manage your personal life even when working hard on a busy firm. Scheduling, having a diary, and being methodical are all essentials to balancing the work/life equation. Put down all your on-calls, night duties and appointments in your new diary as soon as you get them. That way you will not have any embarrassing calls from your Registrar (or even worse, your Consultant) asking you where you are and why you are not at work whilst you are in the middle of a pedicure or in the pub.

In keeping with the competency-based training that is now commonplace for doctors of all training grades, House Officers must provide evidence of achievement of the competencies that are required by the end of the Foundation Years. This takes the form of appraisals, assessments and feedback on your performance by your colleagues. You will be assigned an Educational Supervisor who will oversee your progress throughout Foundation Year 1 and a different Clinical Supervisor within each rotation who will be your mentor through that post.

Most Deaneries have moved away from paper form-filling to a standardised on-line 'e-portfolio' to keep a record of your progress. On starting your job, your Foundation School (usually via the Foundation Years co-ordinator) will provide you with an e-portfolio log in and password. Make sure you familiarise yourself with your e-portfolio as soon as it is available for you to view and make yourself aware of the assessments you need to complete and the deadlines by which you have to complete them. Remember it is your responsibility to arrange meetings with your Clinical and Educational Supervisors, and to arrange completion of your assessments. You will need to use the e-portfolio regularly to record these meetings at the start of your firms, mid-point reviews and end of placement appraisals.

Keeping on top of your e-portfolio (or equivalent) will help keep stress levels low at the end of each rotation and at the end of the year. You will be assessed on your completion of Case based Discussions (CbDs), Directly Observed Procedural Skills (DOPS), mini-Clinical Evaluation Exercises (mini-CEXs), and on formal Multi Source Feedback (MSF) submitted to your Educational Supervisor by the other professionals you work with.

Try to stagger the completion of your assessments throughout the year as opposed to leaving them all until the end. Not only will this help you feel more on top of things, it will give you the time to reflect on the assessments and improve on your weaker points. You should also make an effort to fill out the other parts of the e-portfolio gradually as you progress - in particular, the sections for reflective practice and your personal development plan. An e-portfolio completed in fifteen minutes at the end of May is obviously not going to look good!

Make sure you attend your scheduled teaching. Remember you are a trainee and the teaching is a requirement in your training. This teaching is meant to be bleep-free, so arrange for a colleague to hold your bleep or surrender your bleep at the beginning of the teaching. It is difficult to remember sometimes but the hospital is not going to stop functioning without you for an hour. Make the nursing staff looking after your patients aware that you will be in teaching for an hour and they can bleep you afterwards, or call another member of your team. Be supportive of your colleagues if they need you to hold their bleep as you will need them to return the favour at some point. You will have to sign an attendance register for teaching and will have to account for non-attendance. If you repeatedly find yourself unable to attend teaching, discuss it with your Educational Supervisor so that they are aware and necessary changes can be made.

However do remember that training is not just scheduled teaching; part of training is learning on the job. Ward rounds, seeing new patients, reviewing a sick patient, being on call and so on are all part of your training. Reading up about what you have seen during work will help consolidate your acquisition of knowledge on the job.

Look after yourself by getting enough sleep and eating well. Nothing makes a long ward round more arduous than being exhausted, hungry and thirsty. Make sure to keep hydrated throughout the day, especially during on calls. If you get ill, stay at home - nobody will thank you for spreading your virus about the wards. Give as much notice as you can to your team and Medical Staffing if you are ill so that cover can be arranged. You may have to produce a certificate from your GP if you have been ill and may have to have a "back to work after sickness absence" interview with Occupational Health on your return. It should go without saying, but conversely, do not be tempted to abuse sick leave policy – it will not go unnoticed.

Book holidays and annual leave early – check with your colleagues that your leave does not clash and arrange on-call swaps as necessary. Most Consultants and Medical Staffing will expect at least six weeks' notice so try to plan ahead. Even if you do not feel you need a rest just yet, after some weeks you are likely to, and will be glad of the booked leave as something to look forward to if work is getting a bit tiring. Claim the annual leave you are entitled to, especially if it does not roll over to the next rotation. You will be working hard and you should rest.

Even when not on leave, make some time for yourself to do the things you enjoy. If you find yourself feeling like you have no life outside of work, dedicate a portion of the little free time you do have to things that are important to you. Although you are unlikely to have the luxury of spending so much time pursuing all the extra-curricular activities that you enjoyed as a student, it is worth making an active effort to maintain those that you prioritise. It will mitigate stress and may also show future interviewers that you remain a well-rounded character. There will be times when you have worked twelve days in a row or are in the middle of a stretch of nights when you might feel a bit jaded. Try to remember that working like this is a rite of passage for a doctor and everybody has to do it sooner or later. Working under duress, on-call or on nights is where you are likely to gain most from your experience. It is good to struggle a bit, sometimes. When your on-call or your night stint is over, you will feel more like yourself again (after a good sleep and then spending your free time doing things that are more fun than answering your bleep).

Finally, make the most of your Foundation years as they will form the basis of your training and career; enjoy them! Here are our tips for getting the best out of the personal side of the House Officer year:

1. **Be organised; keep a diary of your commitments.**
2. **Attend any shadowing or induction week.**
3. **Keep on top of your e-portfolio…**
4. **…and attend timetabled teaching.**
5. **Sleep well, keep well hydrated and don't miss meals.**
6. **Book your annual leave in good time.**
7. **Admit when you are ill and stay at home until you are well again.**
8. **Be polite and helpful - this part of being a team player.**
9. **Plan time for yourself, especially after a weekend on call or a stint of nights.**
10. **Enjoy the teamwork in your Foundation years!**

3. What to Write in the Notes

When you leave medical school you will be very familiar with the process of clerking patients and documenting your findings in a conventional format. However, the majority of entries in the notes (and you will be making most of these) are not first-encounter clerkings; they will be progress notes, ward round summaries, records of conversations and on-call reviews.

Most guides on how to document in the notes will start by telling you that the notes are a 'legal document'. Whilst we must appreciate that good quality, accurate documentation is essential, allowing this terminology to raise the spectre of complaints or litigation over students and junior doctors can be counter-productive.

Documentation in the medical notes is an integral part of good patient care. The purpose of this chapter is to explain the *medical* purpose of writing in the notes and our clinical obligations that surround this. Get the 'medical' documentation right and fulfilling any 'legal' obligations will follow.

But what *do* you write in the notes? What information *should* you record and why is it so important to get it right?

Why Do You Write?

The main purpose of medical notes is to record events and management plans together with the information and thought processes that inform those plans. An entry in the notes will be read by a large number of people throughout that patient's stay and beyond. You will read your entry the following day to remind yourself of how things have been progressing, other doctors will read it if they are asked to review the patient on-call and other healthcare professionals (for example nurses, physiotherapists and occupational therapists) will read it to know the thoughts of the medical team. An entry in the notes is therefore more than a brief aide memoire. It serves as a formal record and a means of communicating with colleagues. As such, it is best to write your notes entries with the next person who reads them in mind.

The next professional to read your patient's notes should be left in no doubt as to the information upon which the undertaking of any investigation or treatment to date has been based. If the clinical information and reasoning are clearly laid out for them to understand, they can then proceed accordingly. The other reason for recording this is so that these decisions can be justified if they need to be revisited at a later date.

In addition, the patient and their family will be involved in the management, so any relevant discussion with them forms part of the care and should be documented. This is particularly important since the root cause of most clinical incidents and complaints is poor communication.

When you attend a patient or involve yourself in their management at any time, you assume a duty of care. Even though the Consultant's name is above the bed, this does not absolve a junior doctor (you) of responsibility for those parts of the patient's care they were involved in. You should therefore record your involvement to demonstrate how you have fulfilled this duty.

You are also accountable for your documentation of senior doctors' attendances. Clearly this is a shared responsibility and if you are uncertain what you should be writing then make sure you ask. An increasing awareness of accountability and concerns around litigation means that you will often find that seniors will check what you have written or even write in the notes themselves when an important discussion has taken place or a difficult decision has been made. That said, it is much better if the Consultant comes to write their entry but decides this is not necessary because you have already done it perfectly for them!

The final reason to write in the notes is to acknowledge the limits of your responsibility and demonstrate that senior input has been called upon when appropriate. If you are uncertain or concerned about your patient's clinical situation then you should have no problem in contacting a senior colleague, but it is extremely important to make note of the outcome of this upward referral.

In summary, we write in the notes to serve as a record of events and decisions, of the information that gives rise to them, and to communicate with our colleagues. We also write to document other people's contribution to the care of the patient and to uphold duty of care. The legal consequences and your responsibilities are important, but if you treat the notes primarily as a tool for organising excellent clinical care, you will have fulfilled your legal responsibilities without feeling threatened by them.

When Do You Write?

An entry should be made in the notes whenever a doctor sees a patient or when any change to their management is made. This will therefore include any form of ward round or patient review but also the input of other professionals including telephone advice, verbal reports or results of discussions at multidisciplinary team meetings. Make note of procedures undertaken, of discussion with the family and of any complaints.

Notes should be kept current and any delay in making an entry should be minimal. You must not leave work with any record-keeping pending. Where an entry is made retrospectively, this should be clearly indicated, with an explanation for the delay as appropriate. The current time and the time of the event to which the entry refers should both be recorded. Never go back and tamper with an entry in the notes – it is illegal.

What Do You Write?

Think before you start to write. Gather your thoughts and consider what you are going to say. Much like presenting a case verbally on a ward round or in an exam, if you start and hope inspiration will come to you half way through the presentation you will flounder and get it wrong. Good quality documentation is the written counterpart of good Clinical Presenting.

General Principles for Every Notes Entry

Getting the basics right

Each notes entry you make should start with the date and time (often omitted and *very* important). It should be concluded with your signature, your printed name and bleep number. This means your own bleep number not the on-call baton bleep – pointless if someone needs to discuss something with you later on.

Headline each entry with the context: "Consultant ward round, Dr. X" or "Discussion with Radiologist, Dr Y". You must write in black ink so that any copies made are legible.

Where you make an error in what you are writing, score this through with a single line and note that it was an "error". Do not scribble it out. This may be mistaken for having hidden something if the notes get reviewed later.

Making it reader-friendly

Your notes will be useless unless they are legible. Whilst it is unfortunately still relatively common to have to spend time deciphering some doctors' entries, this is not something for you to emulate. If your colleagues cannot read your entry, it is very unlikely that you will fulfil the objectives of a capable and trusted doctor described in the Introduction to this book. If your handwriting tends to the messy side then you need to print or write in capitals – do whatever it takes to make it legible. Diagrams are often helpful, particularly where a lot of description is required such as in Surgery, or Dermatology.

Have a consistent structure for your ward round entries (see below), and use subheadings. If your entry is clearly signposted, your colleagues will know exactly where to find the information they are looking for.

Making it clear

In order to be clear and effective, your entry should be concise. In the heat of a review of a sick patient, no-one wants to have to read a three page narrative on the minutiae of an entirely irrelevant aspect of the past medical history. A good rule of thumb is to include only the facts that informed your decision-making process.

Writing the salient points in note or bullet point form is perfectly acceptable and usually preferable. Bullet point the patient's current problems at the top of your entry and when it comes to your management plan, bullet point the steps here too.

Do write down your working diagnosis and your differential (see On-Call section below).

The management plan is the area in which you should elaborate on content. This means that the plan should be unambiguous, such that anyone picking up the notes after you would know exactly what has been done, what needs to be done, and what needs reviewing and when.

Language

The language you use must be professional at all times. The two most important things to bear in mind are not to leave anything open to misinterpretation and not to write anything that could be misconstrued as offensive by anybody.

Misinterpretation

It is desirable to be brief in your note-keeping, so there is no need for extensive prose. Note form documentation is ideal. However, if you are going to use abbreviations these should be recognised, and conventional, not your own personal code. If you are using an infrequently used abbreviation, you should also write it out in full the first time it is used. Remember that abbreviations that are familiar to professionals working in one specialty may not be so obvious to those of another. A presentation of "FOOSH" is likely to be puzzling to someone outside the Orthopaedics or A&E worlds where "Fall on out-stretched hand" is so run of the mill that abbreviation is commonplace. Avoid ambiguity.

Avoiding Offence

Under no circumstances should you allow the language you use to be a potential source of complaint. Not only is it unprofessional to use disrespectful or inflammatory language in the notes, it can be very damaging. Always think "Would I be happy for the patient to read this?" and "Would I be prepared for this to be ascribed to me?" In diseases with perceived stigma such as HIV/AIDS you can imagine how your choice of words, if insensitive, could prove distressing to the patient should they see them, and be detrimental to your professional relationship. Your choice of words is also particularly important when writing about delicate situations such as suspected non-accidental injury in children, where no premature ascriptions of guilt should be implied in any way. At the same time, you should clearly record your reasons for concern and do so fully. In cases such as this, stick to facts (your clinical findings) and do not stray into conjecture.

Similarly, the patient notes are not the place to vent your frustrations. Sometimes systems do fail in hospital. Occasionally important pieces of information do get overlooked and investigations or treatments do not always happen when you want them to. Avoid putting down any suggestion of blame in the notes and do not make any accusations when an aspect of patient management has been sub-optimal. Simply document the action that you are taking to put things right.

Writing in the Notes in Specific Contexts

Formal ward rounds (See also Chapter 4 The Ward Round)

As suggested above, you will be the person expected to record all the clinical details, therapeutic decisions and the rationale behind them that arise from a formal ward round.

In the setting of a Consultant Ward Round, when trying to balance the charts and blood results in one hand whilst taking down a wound dressing *and* writing in the notes at the same time with the other… things can get a bit scrappy in the note-keeping department. To add to your confusion, the Consultant may not do things 'in the right order'. They might ask for the results, then examine the patient then get a bit of history and chat to the relatives interspersed with suggestions of a plan whilst you are doing all of the above and trying to keep a logical order to your writing. Having a structure to all your formal ward round entries is the key to getting this right. A good ward round entry in the notes should have five components:

Problems - a summary of the patient's current active problems and any relevant co-morbidities.

Progress - changes in symptoms, any new problems, results of investigations (laboratory results, radiology and so on), the outcome of any requests for advice from other teams and therapeutic interventions undertaken (for example, ERCP outcome or number of days treated with antibiotics).

Examination Findings - record the examination findings in a conventional format.

Impression / Working Diagnoses - overall clinical assessment as a result of the consultation. 'What's going on?'

Plan - well structured and unequivocal so that nothing is missed or open to misinterpretation.

In completing the 'Problems' section properly, you set the scene for every ward round encounter, particularly formal rounds with the Consultant or Registrar. This, together with 'Progess' and 'Examination Findings', justifies what happens next so it is important that it all gets recorded. Some Consultants will examine the patient and helpfully tell you their findings as they go along.

Others will only mumble to themselves if at all. Juniors are often reticent to question their seniors, but in this context, do not be afraid to speak up and ask what they have found "So that you can document it for them".

Leave gaps under each subheading to fill in as the consultation dots about and just strike through any blank spaces if you need to at the end. This maintains a logical lay-out and it is much better than trying to cram things in. Provided you have not consumed half the rainforest in paper in your first month, it will be appreciated.

In order not to leave an on-call team in doubt, difficult decisions regarding resuscitation or "ceiling of care" should be taken on the ward round. Important decisions need to be written clearly and unambiguously in the notes. You may well be the person who prompts the team to consider such issues, and who makes sure they get documented properly.

If you are unsure of something, the ward round is the time to ask, so again, do not be afraid to do so! The chances are that if you are uncertain about why a particular test is being done or how to deliver a certain treatment, then the next person to pick up the notes will be too. Rather sound a little silly in front of your seniors (perhaps avoid doing this in front of the patient!) than miss something significant. More importantly, of course, you do not want to run the risk of compromising the patient's management. The more clearly you spell out instructions in the notes the better.

A similar pattern is useful for your own ward rounds if you are reviewing patients alone.

On-Call (also see Chapter 5 Being On-Call)

When you are asked to review patients on-call, the above scheme can also be used to structure your entry in the notes. You should note that you are the on-call doctor, note why and by whom you have been asked to see the patient, and where you are seeing them.

Summarise the active problems, pertinent components of their medical background and the change that has occurred that has resulted in you being called. Then comes your assessment of the patient (relevant history, examination, any relevant investigations), any emergency treatments that you have instigated and your further management plan.

Particularly when you are on-call, assume responsibility by committing to a working diagnosis by signing your name to it in the notes. (The pressure to give a diagnosis must not, of course, lead you to 'make up' symptoms and signs to corroborate this.) This could seem a bit scary – "What if I'm wrong?"

When you are the first person to review a patient who is unwell or you are the doctor doing the admission clerking, you might not have the results of laboratory or radiological investigations yet. Remember that by the time someone else swoops in later and reviews your diagnosis, they will probably have the benefit of such results and also of the patient's response to your initial management, so do not feel as though you will be judged harshly if your initial impressions are reasonable but not spot-on.

Nevertheless, you are still expected to put down in black and white the diagnosis you have made. The key is to commit to a working diagnosis but to be safe by appreciating (in your differential diagnosis) the serious conditions that require exclusion and the steps you are taking to do this. That way, anyone revisiting your notes can understand at a glance the management decisions you have made, given your clinical assessment.

Integral to being a safe doctor is recognising when you are out of your depth and asking for help sooner rather than later if need be. No one will reprove you for having uncertainties with respect to your patient's management provided they are dealt with in a timely fashion. However, you will quite rightly be hauled over hot coals if you do not bring in senior support early for a patient who is acutely unwell and who you are not equipped to treat on your own. In the notes, you should record the reason for your referral to a senior. Subsequently, document any advice that you have taken and from whom (their name, not just "on-call Registrar"), and write down the plans that you have made for review of the patient either by yourself or by the person that you will ask to see the patient next.

Any time you intervene in a patient's care (increase diuretics, prescribe fluids, give analgesia) this will have an effect on the patient. You should therefore not only use the notes to document the intervention but also your plans to follow up its consequences. The timescale for this review and who does it will depend on clinical urgency. If the patient is stable, it may be a review by the team the next day. If the patient is more unwell the review may need to be in an hour's time. If this is across a shift change, make sure your hand-over is crystal clear and deliver it verbally also.

When on take, your clerking should be as thorough as possible. Do document any relevant negative findings. Do bring to the fore any useful information found from reviewing old notes. (It is a good idea to review old notes or clinic letters before seeing a new patient.) Drug histories are worth a particular mention here. Write down the dose and together with the start date, any new dose changes, drug introductions or withdrawals. Allergies should be documented together with the actual effect that the medication has had in the past. Throat swelling is very different to a bit of nausea, which may be just a side effect!

Aside from the patient's immediate medical problem, it is worth writing down anything else you are concerned about and acting on it. This could include questioning a patient's capacity to make decisions or noting odd family dynamics in a paediatric case. If you write it down make sure you follow it up and document how you have done this.

Procedures

Any time that you undertake a clinical procedure, this should be documented in the notes. This will be important in the event of any complications developing and is also necessary for routine patient management. A simple example is that the length of time a urinary catheter has remained *in situ* should be monitored as this is a risk factor for infection.

Whatever the procedure, you should note the name of the person undertaking it and anyone else who is assisting or supervising. You must document the consent obtained from the patient (even, or perhaps particularly, if this is only verbal), any drugs given, the equipment and technique that is used, any immediate complications and how they have been dealt with and any necessary guidance for subsequent management. For example, this might include wound care instructions or advised timing for drain removal.

Taking Consent

There are very few situations where a House Officer should be solely responsible for taking consent. Usually the person undertaking the procedure should obtain consent but this task can be delegated to someone who has knowledge of the procedure and is competent to explain its nature, its indications and its potential complications. You may, at some point, feel under pressure to obtain consent for complex procedures and may not feel competent to do this. This might occur in the setting of a surgical firm, for example, where you are told to consent for an operation. If you feel out of your depth, you should not put your name on a consent form, and you should ask your senior to be involved. At the same time, do show willing to learn, as a good attitude will demonstrate that you are not simply avoiding responsibility but being aware of your limitations.

Advice

Whenever advice is taken from another professional that will influence your patient's management this should be recorded in full. This may take many forms, for example, advice on antibiotic therapy over the telephone from the on-call Microbiologist or the outcome of a multi-disciplinary meeting between Radiologists, Surgeons and Physicians.

You should note the name, grade and specialty of the person giving the advice and how it was delivered (a telephone discussion, for example). Of course you should note the nature of the advice you have been given but make sure you include the rationale that person has given for their suggestions. Document the changes to management you have made as a result of these suggestions and any further changes or monitoring that need to be carried out and by whom. For example, you may have to ask the night doctor to take a pre-dose Vancomycin level if you have started this drug. Write that you have discussed this with the relevant parties.

Communication with other professionals

It will more often than not fall to you as the junior member of the team to be the go-between in discussions with other professionals involved in patients' care. This will include people based within the hospital such as the Registrar from another specialty, Physiotherapy and Specialist Nurses but also those outside the hospital involved in longer term management such as the patient's General Practitioner, social services and nursing homes. Whenever a discussion is held with another professional about the patient, you should record the issues that have been discussed and the outcome of your conversation so that everyone is kept abreast of progress in their care. If you have asked other people to undertake particular aspects of patient care then document this at the time, especially if it is something important. For example if you have asked the nurse to take sutures out and re-dress a wound then write this down.

Full instructions to the GP and to the patient in the discharge summary are your means of ensuring a safety net for your decisions. Take the time to list fully the patient's problems, results of investigations and how these have been interpreted to initiate treatment in hospital. Clearly state the team's plans for follow-up and which components of this you would like the GP to undertake.

Communication with patients and family

As medical students we learn that it takes time for news (particularly bad) to sink in and information is often not retained at the first delivery. We have all had experiences of patients getting completely the wrong end of the stick despite things having seemingly been fully explained. It is the dubious privilege of the junior doctor to watch a patient nodding vigorously as the Consultant explains their provisional diagnosis and proposed investigations, only to have them whisper, "So what was that all about, then?" as soon as the Consultant (and with him all the answers) disappears off the ward.

Good communication is vital for patients and doctors to work well together and this requires time, clarity and often a lot of patience on your part! You need to establish that the necessary information has been properly received and understood. This is why, in addition to your excellent communication skills, you must record the important discussions you have with your patients and their relatives or carers. It will usually also be your role to record significant conversations your seniors engage in too. Note who in addition to the patient is present, their name and relationship to them and any active involvement they have in the conversation. Record any particular questions or worries the patient has raised and how they have been addressed. This way it is easy to refer back to what was said later on.

For the benefit of the patient, state in the discharge summary the symptoms that should prompt re-attendance at hospital or to the GP's surgery. This is particularly important in primary or emergency care. Your written communication will go home with the patient while your verbal instructions, important as they are, may well be forgotten.

Complaints and when things go wrong

In real life things do not always go to plan. If a family is upset, or if you have encountered a complication with a procedure, for example, the worst thing you can do is try to cover it up. In fact, saying "Sorry" often goes a long way to appeasing patients and their relatives. When documenting an incident, stick to facts and make a note of what has happened, why it has happened (if evident), and what you have done to deal with it. Things do sometimes go wrong and documenting things in this way demonstrates your ability to assume responsibility where appropriate.

Summary

You will find yourself writing in the notes more than any other member of the team. As such, you are the lynchpin of good record-keeping. The purpose of this is to facilitate the best possible medical care.

With this in mind, wherever you are working and whatever the context, make sure that *your* own notes are of the highest quality – it will stand you in good stead for the whole of your career.

4. The Ward Round

The ward round is the cornerstone of managing hospital in-patients. Regular review of patients is essential in terms of monitoring progress, detecting changes that warn of deterioration, and to effecting safe and timely discharge. There are two types of ward round – the junior doctor ward round (your own or your Senior House Officer's) and the senior review (the Registrar or Consultant). Each serves the same purpose but offers different points of view and levels of experience.

As a House Officer you are the person who manages the mechanics of the ward round and it will run smoothly if you complete the relatively simple tasks highlighted in this chapter. If you do this well, the team can spend more time focussing on actually providing care for your patients and, on a personal note, you will earn the trust and respect of your seniors. In fact, your role in this daily event is so important that if things do go wrong; wasting time on the wrong wards, looking for results, having to trawl through sets of notes to identify a key bit of the history, and missing patients off the round, it will reflect badly upon you.

The ward round is not simply a mechanism for generating 'jobs' for you; it is a thoughtful assessment of each patient's clinical state and it should be recorded as such in the notes. Frequently, it is how other healthcare professionals find out what the medical team's diagnoses and plans are and allows them to co-ordinate their activities around your team's thinking in the most efficient way possible.

Your junior hospital jobs will be characterised by an ever-changing set of middle-grade and senior doctors. Frequently you and your Consultant are the only fixed components of the clinical team. Therefore it is incumbent upon you to take the initiative and responsibility for the daily ward round, to ensure that events do not overtake you and that you are able to deliver a ward round that maximises its attention on your patients.

Preparation

Preparing for a ward round is key to its success. Much is made of the difference between Medicine and Surgery in terms of ward practices, but the principles of delivering an effective, comprehensive ward round are the same whether you have a handful of post-operative patients, or have been hit with an enormous medical take.

Making time for preparation is important too, and will depend on the number of patients you have. If you have just inherited 20 new patients from your Consultant being on-call, then the 30 minutes that sufficed yesterday will not be enough today. Teamwork is very important and good outcomes can be achieved through using your common sense to divide labour.

Preparation for a ward round involves assembling a list of all the patients that need to be seen and being up-to-date with their clinical progress. This does not just mean whether their wound is healing well or whether they have stepped down to oral antibiotics. It includes their test results, the outcome of other specialty consults, their progress with occupational and physiotherapy, their social needs and their discharge planning.

Knowledge of almost all of these can be achieved by a neatly and effectively compiled patient list and results folder.

The List

Without a good list, it is impossible to co-ordinate a ward round well. Your first tasks must be to ensure that all the patients who should be under the care of your team are on the list and have not slipped through the net, and that you know where they all are. Patients often get moved to different wards overnight when there is pressure on the bed manager to create spaces and when turnover on admissions wards is particularly high. It is therefore worth telephoning wards where you have out-lying patients before the ward round starts in order to check that they are still where you expect them to be. List your patients by ward and, ideally, by bed number so that you do not jump about the list as you go round and you minimise the risk of missing patients out.

For each patient, you should create columns in which to list their **personal details**, important **co-morbidities** and **presenting complaint**, current **active issues** (which might include dates of procedures, a hospital acquired infection, a failed physiotherapy assessment) and **key results** to note. The content of this last column will vary substantially according to the individual patient, and might be their CRP trend, ultrasound result, or the Respiratory team's opinion. The final column is for outstanding jobs in which you should also leave space for hand-writing in new jobs as you go round.

Your list is a dynamic entity and needs to be updated daily with test results and changes in your patients' clinical conditions. Where there are issues surrounding discharge these should be recorded also, under 'active issues'. New patients that are admitted should be added as you pick them up and sometimes you should keep patients who you have discharged on your list if there are outstanding issues with them (for example referral to a Multidisciplinary Team meeting).

This seems like a lot of information to record on a single table and indeed it is. The trick is to keep it as clutter-free as possible. One or two words or abbreviations as a reminder are adequate. Get rid of the out-dated items as soon as you are able to. House Officers often complain about the 'secretarial' nature of their jobs. However, it takes in-depth understanding to be able to summarise a patient's clinical situation effectively in just a few words.

The list should be saved on the hospital computer system (not a portable memory stick) and in a location that can be accessed daily and by all members of the team, even when you are on holiday! You should then carry a paper copy marked "confidential" with you (and a few spares). Take great care not to misplace it as it contains private information about patients (hence not using portable memory sticks). Old lists should be placed in the 'confidential waste' bin at work and not taken home.

It is very helpful to carry with you the telephone numbers you will frequently need, perhaps as part of the template for your list. You can always call the hospital switchboard, but this wastes a lot of time. You will develop your own 'directory' but a useful start could include:

- **Wards you cover**
- **X-Ray, CT & Ultrasound**
- **Biochemistry, Haematology & Transfusion**
- **Microbiology & Virology**
- **Pharmacy**
- **Bleep numbers for your team**
- **On-call House Officer, SHO and Registrar bleeps**
- **Critical Care Outreach**
- **Nurse Specialist (if relevant to your job)**
- **Physiotherapists & Occupational Therapists**
- **Social Workers**
- **MDT co-ordinators**
- **Discharge co-ordinators**

The Folder

Most House Officers find it helpful to carry a folder with them (particularly on the Consultant ward round) so that all important paperwork and data such as investigation results are easily at hand.

Blood tests have the potential to become the bane of your working life, particularly when already over-stretched phlebotomy services break down. You will doubtless put all your blood request forms out in the evening and expect that the following day there will be corresponding results on the computer. Unfortunately this is not reliable. The number of steps that can go wrong and thwart your attempt to have an up-to-date set of results on the Consultant ward round is enormous. Our advice is simple – assume nothing!

Develop a healthy neurosis in checking that things have actually been done and do not just assume that the system works. This will make your co-ordination of the progress of your patients' investigations and treatment more efficient. It is your responsibility to act upon a patient's blood results not only to have requested the test, so on occasion it may well fall to you to take a set of bloods yourself. Of course, this does not only apply to blood results but is also relevant to getting x-rays done, dressings changed, medications ordered and so on. Did the nurse actually do the urine dip you asked for? Has the ultrasound request actually been processed, not just sat on the desk of the receptionist? Keep a record in the notes of tests that you have requested so that you can track them and make a note of the result when available. This will also prevent colleagues inadvertently doubling up on jobs. Make sure you chase things up until you are sure that you are going to get results.

The folder should have a flowchart for the blood results of each patient on the firm. You should have a standard, reproducible chart of commonly requested blood tests in a logical sequence and leave space at the bottom to fill in more infrequently performed tests when necessary. Filling the flowchart in by hand may seem a little archaic but it is the only way that you can ensure you will be aware of every abnormal result. Having written down a potassium level of 6.5 should guarantee that you will think that "That's a little high" and you will consider what to do about it! Not only that, but a filled-in bloods folder saves the inevitable queuing for a computer mid-round and speeds up the whole process. Similarly, you should have printed copies of patients' radiology reports where possible. The folder may also be used for keeping useful stationery such as history sheets, 'Not for Resuscitation' forms, consent forms, laboratory request forms and request forms for other tests (for instance cardiac physiology and endoscopy forms). These can then be pulled out, filled in and signed on the spot which saves doubling back later and minimises omission.

The Consultant Ward Round

You will know in advance where and when to meet the Consultant and if you have most of your patients on one ward then you should make sure all the notes and bedside charts for you patients are present before they arrive. There is a large variety of healthcare professionals who may also be seeing your patients concurrently, so do not be surprised if all the notes are not where they should be when you arrive. The ward clerk is often the most useful person to ask if you have trouble locating notes. Many wards have the notes contained in portable trolleys. It can be useful to take the whole trolley on the round with you.

The pace and focus of ward rounds will differ with each Consultant. Find out how your Consultant likes to run their ward rounds; for example whether they want to hear the history before you reach the patient's bedside, or at the bedside.

If the patient is new to the Consultant, a more detailed account of their history will be needed. If the Consultant is already familiar with the patient, a brief synopsis of events since they last saw them will suffice. Know your patients well and be confident in the delivery of information. It is neither necessary nor desirable to memorise every single blood result (that is why you have the folder) but you do need to have a grasp of what has been done for the patient and the plan in place. This is especially important if the Consultant has asked you to request certain tests. If they have not been done, or not requested, then you must be able to explain why.

There is always a lot happening at once on a Consultant ward round, and there may well be more than one junior doctor on the round. Share the roles between you – one can write in the notes while the other presents. Throughout the round, other healthcare professionals may contribute their progress reports and raise their concerns, medical students may get some teaching and the junior doctors may get a bit of a grilling. All of this is excellent learning for you and you should do your best not to be scurrying off, thinking these are 'pauses'. Once finished, ensure the next set of notes is ready and prepare for the next patient, making sure that between yourself and your team-mates the jobs for the patient have been written on your list.

The advice presented in this chapter will make the round more effective, but be prepared for things to not go quite as planned - the above information will not apply for every event. For instance, you may be on your own with the Consultant. However, if you have made the effort to prepare and are organised despite having to do this single-handedly, not only will this be noticed and appreciated, it would be a very hard-hearted senior colleague who does not 'muck in' and help out.

The House Officer Ward Round

It can be difficult for junior doctors to grasp the essence of what is expected of them when consulting patients on their own ward round. We are taught throughout medical school how to take histories, how to examine patients, and how to assess those who are acutely unwell. Rarely are we taught how to approach a patient on a ward round. Documenting what a Consultant says on a round is relatively logical and has been dealt with already (see above, and Chapter 3 What to Write in the

Notes), but how do you approach the patient who has landed on your ward without re-clerking them from the very beginning?

Actually it is remarkably straightforward. The same principles of history and examination followed by investigations should guide you but, with practice, you can tailor your history and examination to the needs of your patient. Quickly read all previous entries in the notes (taking particular note of the admission clerking), read the latest clinic letters (which usually lurk at the back of the notes folder and provide a very useful summary of the background information) and have a look at blood tests and radiology before you go to see the patient. This is the art of reviewing patients rapidly and will be of use both here on the wards and in the setting of an urgent review.

You will go to your patient's bedside knowing their background and progress to date, allowing your interaction with them to be more focussed and more productive for both of you. Ask about the patient's symptoms and now, in a more directed fashion, you can elicit any pieces of information you may be lacking. Examine the relevant systems and make sure you look at the information from the observations, blood sugar, fluid balance, nutrition and stool charts. Act on any abnormalities. See where you are in the list of jobs from the previous senior ward round. Keep your own job list and note down queries you may have for your seniors. By the end of the whole process if you can leave your patient's bedside knowing the answer to the 5 simple questions that follow, you will have a good appreciation of your patient's clinical situation:

What have they got?

What haven't they got?

Why have they got it?

What are the complications?

What else is going on that might affect my management plan?

Irrespective of whether the patient has been clerked by your third year medical student or fully worked up by the Acute Care Team prior to arriving on your ward, you should keep an open mind about your patient's diagnosis. Sometimes 'What they haven't got' is more important than 'What they have got'. Clearly, ruling out myocardial infarction and pulmonary embolus is more important that diagnosing musculoskeletal chest pain. Similarly you should be 'hungry' for clinical information – what risk factors does your patient have and what complications (of the primary pathology and of their treatment) have they developed? Also, you should keep a wide-angled view of your patient's condition. Somebody with severe osteoarthritis of the knee cannot be expected to tolerate walking on a treadmill for long so arranging an exercise tolerance test is not a sensible management plan.

The last question is also relevant to discharge, which is an extremely important part of maintaining flow through the hospital. Most institutions work either at or around capacity at all times. Delays in discharge have significant knock-on effects to the running of the hospital, to the scope for elective procedures to take place and they place the patient concerned at unnecessarily increased risks of acquiring nosocomial infections. As the junior doctor, and therefore the team's representative to the ward's Multidisciplinary Team (see Chapter 1 Making Ward Life Work), it is your responsibility to appreciate the social set-up of your patients and any barriers to their discharge.

Once your assessment of the patient is complete you must document your encounter with the patient in their notes. Legally the team is obliged to write in the notes every 72 hours, but it is good practice to do so daily. It can be challenging to document your thinking and findings concisely yet completely. How to go about doing this is tackled in Chapter 3 What to Write in the Notes.

After the Round

Things are not quite over yet. The list of jobs you will have generated after the round need to be reviewed to make sure you have not missed anything out and that each task is allocated to a member of the team. This may be you and your counterpart House Officer or it may involve one of the more senior members of the team. There is no right or wrong way to go about this. However, it *would* be wrong not to use this time to reflect on your preparation, on the round itself and on the tasks ahead of you to ensure that you grasp any learning points. If you are designated particular investigations or specialty opinions to request, then now is a good time (if you did not already do so at the time they were proposed) to make sure that you understand exactly why they are going to be useful and what clinical question you are trying to answer. This will ensure that you make these requests appropriately and is dealt with further in Chapter 6 Referrals, Requests & Tests.

Summary

The ward round is where continuing assessment and decisions are made. Its success depends on effective preparation beforehand, an organised approach during and appropriate division of jobs afterwards – with a healthy obsession about getting these jobs done. It is an unthinking House Officer who believes they are 'just a secretary'. This is your chance to showcase your development as a junior doctor to your Consultant, so make sure you understand your role in the ward round and fulfill that role to the best of your ability.

5. Being On-Call

When you are on-call you will be a member of one of two teams. The ward cover team exists to ensure in-patient safety out of hours. The admitting team assesses patients referred by General Practitioners or Casualty and initiates emergency and continuing care.

Clerking / On Take

The Admissions Unit

Larger specialities like Medicine and Surgery will usually have a dedicated admitting team. In the morning, you will generally collect the bleeps and receive handover from the previous team, and may also form part of the preceding night's post-take ward round.

Hospitals differ in their admitting policies. Most have Medical Admissions Units (MAUs), or combined Medical and Surgical Units, whilst others may admit directly to a ward. The dedicated units are generally well organised and equipped, often with a dedicated Consultant, Registrar and an SHO. Whether the MAU is staffed by a dedicated set of doctors or by a rotation of doctors from ward-based teams, the 'day team' will usually consist of a combination of grades of doctors on-call. Referrals from Casualty or GPs will be directed to the Registrar or to the SHO. The accepted names will be added to a 'take' list and you will be then be expected to clerk people in the order in which they arrive. Depending on clinical need, more unwell patients may get 'bumped up' the list.

In some hospitals, the House Officer will carry the external referral bleep for GPs, although this is becoming less common. When taking referrals be polite and listen to the GP's story. They are likely to be as busy as you, and will certainly be more experienced. It is not your job to bounce GP referrals nor to give them a hard time. You are speaking to a senior doctor who does not have access to any of the tests you take for granted within the hospital and who is concerned enough to call you. If the GP wants to send the patient to hospital for an opinion from secondary care, it is very difficult to refuse (and probably wrong to do so). However, whether the patient is best served by being seen by your team as an emergency or in the outpatient setting or perhaps by another team in the hospital are important considerations, and in the event of any uncertainty you should discuss with your Registrar and then call the GP back.

It is worth remembering that the GP is likely to be within earshot of the patient either in their home or in the surgery. Take down the patient's name, date of birth, and details of the presenting complaint. Advise the GP to send the patient to the Admissions Unit, and politely ask that they send a referral letter with the patient, preferably with a past medical history and list of current medications including any recent changes. You will then need to speak with your Nurse Coordinator to let them know about the referrals so they can organise bed spaces.

Some hospitals like all patients to be seen in Casualty. This can be difficult for a number of reasons. The time constraint imposed by governmental waiting time target means you will be pressured into making a decision to admit or discharge. Most hospitals will have a policy that precludes House Officers from discharging patients independently so do err on the side of caution and admit people, if only for a senior review later. This will minimise error and protect patient safety. Nurses in Casualty are generally excellent, but very busy. This can make setting up equipment and assistance for more detailed examination and procedures tricky - another good reason for admitting. Patient safety should always trump bed management, governmental or staffing issues.

Although your patients will have already been seen by the Casualty team or the GP, you will be the first doctor who can afford to spend time taking a thorough history and performing a detailed examination. You will be slow at first, but practice consistently and efficiency will follow.

If you are concerned about a patient, do not wait until the formal ward round. Some Registrars and Consultants will review patients as they are admitted, but if not and you are concerned, ask for help. It is a strength not a weakness. It can sometimes be difficult to walk the line between seeking appropriate help and having the confidence to start basic investigations and treatments before asking your senior for their support. This is something which you will have to figure out early on in your career. The best approach is to be enthusiastic, take on board people's guidance, and grab every opportunity you can to build up your skills and your confidence so that you can be a valuable part of your team.

Unwell patients can and will deteriorate quickly. The most important part of your clerking is a clear plan that the nurses can follow. After you have finished your writing, seek out the nurse who is looking after the patient and explain your plan and what you would like them to do straight away. This may be catheterising the patient, taking a urine sample or giving intravenous antibiotics. This optimises patient care and develops a good relationship between medical and nursing staff.

Post Take Ward Round

A good House Officer should be able to conduct and lead efficient ward rounds and handovers. The timings of post take ward rounds will vary between hospitals. On a medical take, there is usually an afternoon or evening ward round to see patients admitted from 9-5, and then a morning round to see patients from the evening and night takes. It will be your responsibility to create a ward round list and this needs to be accurate. There may be a template available on your MAU computers, but if not, draw a simple table in Microsoft Word which clearly states the Consultant's name and the date at the top. For each patient the table should contain their full **name**, **date of birth**, **hospital number**, **ward** and **bed number**, **presenting complaint**, **doctor** who clerked them, working **diagnosis** and a column for **jobs**.

Conducting the post-take ward round is much like running the Consultant ward round (Chapter 4 The Ward Round). However, it is more likely in the post-take situation that someone has been identified as being particularly unwell, therefore your 'route' should accommodate this. If there is someone who is likely to become more unwell over the next few hours and particularly if this means overnight, ensure with your seniors that there is plan documented in the notes and that the nursing staff and relevant medical teams are aware (see section on handover).

Crash Bleeps

The Medical Day Team bleeps usually double as cardiac arrest bleeps. The Cardiac Arrest Team, usually consists of the Anaesthetist on call, Medical Registrar, SHOs, House Officers, and Resuscitation Nurses. Depending on the site of the arrest, you may be the first person there. It is acceptable to feel nervous, but do keep calm and go back to the basics. You will be trained in ILS (Intermediate Life Support) by the end of medical school and you should aim to complete an Advanced Life Support course as soon as possible (www.resus.org.uk). At the arrest, your colleagues will know that House Officers are inexperienced and not expect unreasonable heroics from you, so make the most of this time to learn and participate actively in as many arrest situations as you can.

Ward Cover

The prospect of ward cover usually strikes fear into the heart of every junior doctor. It need not. Prioritise your work, stay calm and make the most of this unique training opportunity.

As the doctor covering the wards, your role is to treat patients who become unwell unexpectedly and to facilitate patients' ongoing care according to the plans of their regular teams. The latter may involve planned tasks that have been handed over to you, such as taking blood at the appropriate time for drug level monitoring, or undertaking reviews of patients whose clinical condition is known to be unstable. It may also include unforeseen jobs that the nurses request of you such as re-cannulating a patient. Every job that you do should be done thoroughly. For example, when prescribing intravenous fluids, ensure that you know why the patient needs them, assess their fluid balance clinically and check the latest urea and electrolytes.

Weekday ward covers are generally quieter and from Sunday to Thursday you know that the regular team will be available the next morning, so non-urgent jobs can potentially wait until then. Weekend ward covers will be busy, but you will learn the most during these on-calls.

After the bleep and any outstanding jobs are handed to you by the regular team, go to the wards you are covering. Introduce yourself and request that the nurses stack up non-urgent jobs and bleep you for all the jobs together. If they can put a friendly, approachable face to the title of 'doctor on call' nurses keep calmer and the torrent of unnecessary bleeps is reduced.

For the next few hours, any bleeps that come to you should then be regarding a patient that the nurses are concerned about. The Modified Early Warning Score (MEWS) and other similar scoring systems give a score to each parameter of the vital signs and are calculated for each patient every time a nurse makes their observations round. The nurses have to inform a doctor if the MEWS reaches a certain number (determined by your hospital or adjusted by the regular team to suit a particular patient). You are then expected to review the patient within a given time frame. When you answer the phone, bear in mind that the nurses calling you are just as busy as you are, and have no idea what else you may have been up to when they bleeped. Your job is to be 'on-call' and they are calling! They will often quote the MEWS. You need to ask which parameters are abnormal in an individual patient's case, not just the overall score. Information that you need from the nurses is that which will allow you to prioritise your review of that particular patient. Request that the nurses stack the observations charts, drug charts and the patient's current and old notes on the nurses' station in readiness for your review, as this will save you precious time when you arrive on the ward.

Depending on the reason for the call, there may be some early interventions that the nurses can undertake before you get there. If the patient has low oxygen saturations, for example, you could ask them to give supplemental oxygen. If you feel the situation has to wait until you have attended to a patient who is more unwell, explain to the nurses that you will be there as soon as you can.

Sometimes, you will be called because the patient 'doesn't look right'. Although less objective, this is frequently the most worrying call – experienced nurses tend to have gut instincts that are often correct. Request a set of vital signs measurements and attend to the patient.

On arrival, seek out the nurse looking after the patient – ideally they should be present when you are undertaking your review. If the patient looks unwell, go back to your basic life support training (ensure a patent airway, oxygenation and circulation) and perform a quick focussed history and examination. Seek help early. Your Registrar will want to know that there is a sick patient in the hospital and can advise you on further management. If the patient is alert and oriented, you have time to turn to the notes and gather some more background information.

In your notes entry in this context, it is particularly important to document the time at which you have attended the patient and the ward on which you have seen them. Document the reason for which you have been asked to attend and then make a quick summary of the presenting complaints and important past medical history. Is this someone who is normally fit and mobile or someone who is dependent for all activities of daily living, in a nursing home? Is there a plan documented in the notes by the day team for reference in the event of further deterioration? Has a ceiling of management been agreed with the patient, the family and the staff? Is there an existing 'Do Not Attempt Resuscitation' (DNAR) form? Be aware that not attempting resuscitation is very different to not giving full and active treatment.

Complete your assessment, make your initial plan and inform the nurse. Reassure the patient by letting them know that you have instituted a management plan and that the outcome of this will be reviewed. As appropriate, request the investigations you need - take bloods, arterial blood gasses, request a chest radiograph or ECG. Ensure that you return to review the patient and any outstanding results. Give a time frame for doing this and tell the nurses. If you have given them an estimated time for your return this will stave off any unwarranted bleeps in the interim unless truly urgent. It is then important that you do not undermine their trust and do come back when you said you would. If another emergency keeps you away, give them a call and tell them (or ask someone else to do this for you). If the patient is unwell, do not leave for less important jobs. There are other people in the hospital capable of writing up fluids, cannulating and re-boarding drug charts for patients elsewhere, including those senior to you - unwell patients come first.

Nights

House Officer night shifts are slowly dying out, to the detriment of training, team working and the annoyance of the SHOs who have to cover instead! Nevertheless, many hospitals still have House Officers on night duty. The same rules apply: prioritise, keep cool and look after yourself.

For some specialities you may be the only doctor on the shop floor, covering admissions and the wards. If you need help, do not hesitate to get your Registrar out of bed, they are just as on call as you are!

Everyone has their own way of handling the transition to nocturnal living and of surviving the 'jet lag' feeling during nights. However you choose to 'get through them', after finishing nights do not drive long distances straight away, as the adrenaline is likely to wear off halfway through and you are then at high risk of serious accidents.

Handover

Effective clinical handover relies on good interprofessional communication skills. There are many situations where you hand over patient details and management plans to your colleagues both formally and informally.

A post-take ward round is essentially a handover of the care of a patient who has been clerked to the team who will be responsible for them during the remainder of their hospital stay. This handover can take a number of forms. For example, in most Orthopaedics units, there is a morning trauma meeting to review the histories and radiographs of patients admitted the previous day, prior to the ward round.

Most hospitals operate a written handover system for the weekends. The regular teams will highlight a list of jobs that the weekend ward cover doctor needs to carry out, or patients who might need review.

When you finish your on-call, hand over any patients that you think need a review directly to the doctor taking over from you, and if appropriate you should write that you have done this in the patient's notes.

Summary

Being on-call is a big responsibility and a challenging learning opportunity. However it is during these periods that you will have the most memorable educational and professional experiences. You will meet and work alongside numerous doctors from across the hospital, often in pressured environments, thereby developing your own network of contacts.

Draw upon all your skills – clinical, personal and organisational – and you will make the stress of the on-call worthwhile.

6. Referrals, Requests & Tests

One of the most challenging parts of the job of a House Officer is making referrals and organising tests. These tasks involve conveying information to another doctor or team – often to a person with more experience than you and it is not unusual to feel a little daunted. While making referrals well is satisfying and productive, making referrals or requesting tests badly results in the creation of more work, as the person at the other end asks you to 'get more information' or 'find out from your Consultant why they want this task done'. This in turn means you have more to do before calling them back, wasting your time and their time – a frustrating process all round that simply serves to delay patient care. This chapter will give you some basic handy hints as to how to make a great referral or request so your patients get the best and most speedy service possible and of course making you look good!

Who to refer to and when to refer?

If you are trying to organise investigations or require an urgent (same day) opinion, you must make your request as early as possible in the day, giving others the chance to structure their time.

The person you are calling will more than likely have a full day of work ahead and will have to come and see your patient in addition to their own commitments (theatre lists, outpatients, ward rounds and so on). They will be able to plan to see your patient more easily if you call them early.

If you leave the request until later on in the day (or even after hours) it is much less likely that your colleague to whom you are making the referral will be able to rearrange their own duties to attend to your patient. The rule here is to make the referral as soon as possible after you know you need a test or opinion. If referrals are made after hours, then you will need to exercise some judgement as to whether or not they truly are urgent. It is absolutely appropriate to make an urgent referral at any time if the case warrants it, however, a non-urgent referral is best kept within hours. If you have any doubt about how urgent a referral is, you are always safer asking your own seniors or even ringing the 'on-call' for advice.

Preparing to make a referral.

Prior to making your referral make sure you understand:

1) Why you are making the referral – what clinical question are you asking?
2) How urgent it is – same week, same day, now?
3) What you want from the referral – an opinion, the patient taken over, telephone advice?
4) The details of the case – the history, the examination findings, latest observations and investigation results.

Being able to communicate this information to the referee will enable them to prioritise your case amongst others and they may then suggest further work-up that might be helpful prior to their review of your patient so that this in turn can be more productive.

Making the call

Find out the correct method of making the referral - a bleep, paper referral, on-line request, letter in the notes or a combination of the above. This information can usually be obtained from the relevant department's secretary, or from one of the junior members of their team.

Gather the required information for each case. If referring several patients to the same person it will be necessary to gather all the appropriate notes together.

If making a telephone referral - call the appropriate extension or bleep. Do not bleep several times in quick succession. Your referee will appreciate your patience and you can bleep again if a suitable time has elapsed without a reply. If you are still without response you can try another member of the same team or the Consultant's secretary to work out how best to get hold of the appropriate person (see 'bleep etiquette' below). Have some additional work with you so that you can get on with something else while bleeping, for example ordering the following day's bloods or doing discharge summaries.

Introduce yourself and mention your grade and team. This way the person taking the call knows what information it is reasonable to expect you to know. If it is your Consultant who has specifically asked for something, say that early – it is useful to know who has originated the request!

State what you are hoping to get from the interaction. For example, "We were hoping you would see this patient with a view to taking over their care", "I am phoning for telephone advice".

Requesting urgent help. There are some circumstances in which urgent help is required – do not hesitate to say if you are concerned about a particular situation and feel out of your depth. For example, if calling the Medical Registrar on call about a patient, "I am a House Officer, I am concerned that this patient is extremely unwell and may arrest and I have no other members of my team available, are you able to come and help?" In this situation however, the above requirements for information still apply as the person responding will still want to know about the case and may be able to guide you through the immediate steps to take while they make their way to join you.

Explain the details of the case and answer any questions.

Clarify the details of the advice you are given and the action that will be taken by the person responding. Will they review your patient today or in two days' time on the Consultant ward round? Will the advice they give you by telephone be the limit of their input unless circumstances change again?

Take the name of the person responding.

It sometimes feels that by modifying the clinical details a little in the 'wrong' direction, it might be easier to get the referral made or test done (for example suggesting that symptoms are more marked than is actually the case). While this might generate more speedy action by the department or person you are referring to, it is **never** the right thing to do. In the first instance, an exaggeration of symptoms may result in a course of action being recommended that is too aggressive for the patient concerned. Secondly, it may give the patient a wrong level of priority which may mean that someone else who is more unwell will be seen later or have their test deferred. Thirdly, while it is entirely appropriate for you to call for help in *any* case that you are genuinely concerned about, fabricating symptoms and signs casts doubt over your own probity and simply means that you are not able to judge your patient's condition appropriately.

After the call

Immediately after your conversation, document it all clearly in the notes. Include the date, time, who responded, what they advised, when they were coming to see the patient, what you were required to do (for example, organise a test and call them with the results). In your documentation thank them for their advice and action. This should particularly be the case if they are going out their way to help.

Bleep Etiquette

- Call switch board to get the relevant number(s) for the team or department you wish to speak to.
- Have all patient information to hand including notes, observations chart, drug chart and investigation results.
- Have other work to hand to be getting on with during the inevitable waiting.
- Bleep once.
- Wait at least 5 minutes.
- Bleep again – do not bleep in quick succession. Serial bleeping does not help.
- Wait 5 minutes.
- Try another member of the team (if only to confirm that you've been trying the correct number).
- Ask switchboard to bleep for you.
- Work your way though all members of the team before calling the secretary and either ascertaining the whereabouts of team members or leaving a message.

Organising tests and investigations

A majority of the tips outlined in the first section of this chapter also apply when requesting tests, but there are a few additional points of note. The background is the same; make sure you know exactly what is required and the reason why (see 'Preparation' above). In a multi-agency organisation like a hospital there is ample opportunity for things to go wrong. An important transferable skill that the house job year should teach you is to be able to consider a process, identify points along it where mistakes are at risk of occurring and pre-empt them. This might be something as simple as thinking through the preparation that a patient undergoing an upper GI endoscopy requires and making sure that the nursing staff do not allow a patient to miss their slot on the list because they have not been fasted properly.

The request form

When trying to organise a test, the first step is filling out the request form. Many of these are badly designed and the boxes are densely packed but there is usually just enough space to get the necessary information across. Always make sure that the following points are thought through and that all required information is present and legible. This will include:

1) The patient's details.
2) The patient's location.
3) The Consultant in charge.
4) Contact details for someone available to discuss the case (not an on-call bleep).
5) The reason for the request.
6) Times when the patient will not be available (because they are having another test or in theatre or dialysis, for example).
7) Is there any infection risk?
8) Patient risks (for example, anticoagulation, contrast allergy, degree of renal impairment).
9) Can the patient consent to the procedure?

Not all of this information is required in every case but all these issues should be considered.

The form will need to be validated (usually by a clinician in that department – a Radiologist, for example) and then given to the member of staff who actually books the test. If the request is submitted electronically, make sure you know where it will be printed out. Going back to the principle of identifying weak points in the chain, if your request prints out at the wrong desk, wrong department or even in the wrong building then it does not matter how good or timely your request was - the test will not happen. It is always a good idea to follow up the request with a phone call later in the day "to find out when the test or procedure has been listed for". This will give you an idea of the waiting time, but is also a way of making sure that the request has been properly received and acted upon.

Requesting urgent tests

The urgency of a particular test most frequently relates to clinical need. Some tests are part of the urgent assessment of a patient's condition and govern the escalation of their care - to a high dependency or intensive care setting, for example. Some tests and investigations have a therapeutic element that will dramatically alter a patient's immediate prognosis and constitute part of the management of the case, for example primary angioplasty or emergency endoscopy.

Whether or not a test is urgent should also take into account how the result will alter subsequent management. A chest radiograph is unlikely to diagnose a pulmonary embolus in itself, but is needed urgently to rule out other causes of the patient's symptoms and may direct further investigation (for example nuclear medicine or computed tomography (CT)).

There are also other more practical factors that will determine the timing of a test. For example, a greater degree of urgency may be ascribed to a test upon whose result a patient's discharge depends. An outpatient appointment may have to be timed so that the results of the test can be discussed in the clinic.

Getting the Test Done

Having requested the investigation, you will need to consider the practicalities of actually getting it done. Your main responsibility is for inpatients, but you will sometimes have to organise tests for patients who have left the hospital. In this case, you will have to consider whether the patient can manage the pre-procedure period as an outpatient or whether they require another inpatient stay for the test?

Think through the process the patient is about to go through from start to finish:

1) Let the ward know what is to happen so that any necessary arrangements can be made.

2) Can the patient travel to the test? (Will your inpatient travel in a bed or a chair? Will an outpatient need transport?)
3) Does the patient have additional needs? (Do they require oxygen?)
4) Can the patient consent to the test? (Do they not speak English and require an interpreter? Do they have dementia and lack the capacity to consent?)
5) Is the patient appropriately prepared for the test? (Have they been nil by mouth for six hours prior to their upper gastrointestinal endoscopy?)
6) Is the patient on any medication that will preclude the test at a particular time? (Are they on warfarin or clopidogrel?)
7) Does the patient have any conditions that will make a test impractical? (Are they unable to lie flat?)
8) Is the patient properly prepared? (Has a group and save been sent? Have a pre-procedure full blood count and clotting been sent? Has the patient had the pre-procedure antibiotics? Is a venflon sited?)
9) Does the patient have any conditions that represent a hazard for those completing the test or patients following them on the list? (Does the patient have any blood-borne viral infections, and should those dealing with the

case be particularly vigilant for body fluid spills? Does the patient have *C. difficile* or MRSA and therefore need to be booked onto the end of the list to allow for deep cleaning prior to the next case?)

10) If a test is being done to obtain a tissue or fluid sample, is the appropriate transport medium available for that sample, and are the laboratory staff aware that they are to receive this sample? You may not know what specific media are required or the protocols for handling specimens, so ask.

Knowing about these issues will allow your patient to have the test they need as soon as possible, without any 'false starts'.

Requesting mobile tests

Some tests can be done at the bedside if the patient is too sick to move. By and large this refers to chest radiographs but can on occasion include other tests such as bedside echocardiography or ultrasound, for example.

If your patient needs it, and you ask for a 'portable' test, it can be arranged. Remember, however, that a request for a mobile test will always require discussion with the relevant department as well as submitting the appropriate form. There are a number of tests that simply cannot be done outside the appropriate area, such as abdominal x-rays.

Seeking advice about test requests

Always remember that those performing the test are experts. They are not just there to do your team's bidding but also have an in-depth knowledge of the tests themselves. They can advise you if you are requesting the wrong test to get the information or result that you require, whether a different test would provide more information and which tests are indicated or contraindicated. This is why it is important to make the clinical question you are asking clear on your request form. If you have any doubt, it is always better to discuss the test and ask advice rather than plough ahead and expose the patient to unnecessary risk. Although it means more leg-work for you, actually going to the relevant department (notes in hand, if need be) to discuss the patient rather than making a telephone call can be a good way of building a working relationship with the specialist concerned and can be more productive. By discussing a test, you will find that the people you ask advice from will help expedite matters according to their judgment of your patient's clinical need.

Summary

Making requests and referrals is not difficult but it is often done badly. If you do this well you will save yourself time and energy and the running of the firm will move smoothly. Experience will change how you make requests with time but this chapter will help you get started.

Making referrals and requesting tests

Do request tests and opinions early in the day.
Do be courteous.
Do know your case.
Do understand why you are making the referral or request.
Do prioritise requests and referrals.
Do ask for advice.
Do be honest.

Don't treat your colleagues simply as service providers – they often know more than you.
Don't bother others if you are not ready to refer. Saying 'I've got one coming your way' only serves to irritate.
Don't try and bulldoze your way through a situation – it may work on that occasion but you will make people less inclined to go out of their way to help you in the future.
Don't lie to ensure the test gets done – this is not in the best interests of your patient.
Don't serially bleep.

7. Audit & Research

Audit

Audit tests performance against a known standard, implements changes in practice to improve performance and then re-tests to demonstrate the effect of those implementations. Audit is 'the systematic, critical analysis of the quality of medical care, including the procedures used for diagnosis and treatment, the use of resources, and the resulting outcome and quality of life for the patient' (Department of Health *Working for Patients*, 1989).

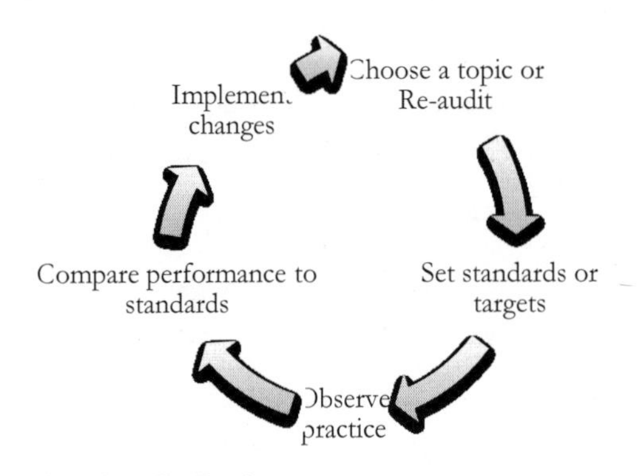

The Audit Cycle

Why is Audit important?

The reasons for conducting an audit or getting involved in audit as a junior doctor are myriad.

Yes, audit fills the gap that should be part of every health care professional's CV. When attending an interview for your next post it would not reflect very well if you had failed to do any sort of audit in your last post. An audit per post or every six months is a reasonable rate to aim for.

More importantly, every health care professional has an obligation to test their performance and specific aspects of their patient care so as to ensure that they are optimal.

Audit stimulates you to ask questions about your practice, to justify your actions and those of other health care professionals, and to modify your performance when necessary. All too often in clinical medicine junior doctors carry out tasks without questioning why they are doing them or why they are doing them in that particular way. Retaining an inquisitive perspective on your profession in the face of a busy house job with an endless and at times a seemingly unstimulating 'to-do' list, is to be applauded.

One successfully completed project can also be the stepping-stone for further audits and also into a research project or study.

Audit helps to integrate a new junior doctor with other specialty teams in the hospital and affiliate them with other health care professionals. Career progression depends on your ability to network, so making yourself 'known' is invaluable.

How do you start it?

A useful audit will ask a question that is significant to the delivery of patient care. The topic may be related to patient safety, working practices or workload. Review of a clinical or serious untoward incident may identify a need for audit. You may even consider the appropriate use of resources, high cost interventions or minimising waste as topics for audit. There should be good evidence available to inform standards - usually systematic reviews or national clinical guidelines. The problem should be amenable to change and sustainable improvement should be possible.

Examples of typical clinical audits

Type of Audit	*Example*
Documentation	Quality of information in discharge summaries
Clinical Intervention	Urethral catheterisation in Care of the Elderly wards.
Diagnostic Accuracy	Retrospective study looking at A&E radiographs and noting all the missed fractures.

If you still do not have a good topic for an audit after racking your brains then ask the Registrar or Consultant on the team for some suggestions. You can easily get involved in an existing audit; plenty of senior doctors are looking for someone to take the lead role on their ideas. The most important aspect is to choose a good question. In general, the simpler the question posed, the easier the audit becomes, meaning your findings have a greater chance of impacting upon your clinical work. Too many junior doctors get bogged down in trying to think of complex problems and dreaming of creating revolutionary change in healthcare. Whilst having high aspirations is all well and good, in real terms, in order for you to complete a productive audit as a junior doctor you ought to look at a study that can be turned around in four to six months. It is therefore sensible to chart a timeline.

Week 1	
Define the topic/question. Identify the standard/guideline.	Draft a proforma and get it reviewed by a senior doctor

Weeks 2 and 3			
Define the study population.	Contact the audit office and register.	Request that they collect the notes.	Start data collection

Weeks 4, 5 and 6		
Analyse data, write up.	Propose changes and implementation strategy.	Present your Audit.

Beyond Week 6	
Carry out implementation strategy.	Plan for re-audit.

Design an appropriate method of selection for the subjects in your study, define your inclusion and exclusion criteria and calculate the sample size (you may need statistical help for this).

How do you conduct it?

Measuring performance or observing practice is the most labour-intensive part of the audit process. Design a form on which you will record all the data parameters you want to collect. Within reason, collecting too much information is not a bad thing, as you can always distil it later. Too little information will necessitate the arduous process of data collection all over again. You may need to register your project with your Trust's clinical audit department who are likely to prove very useful in co-ordinating the study. If your data collection is retrospective and you require patient notes, the coding department will obtain a list of the clinical codes that fit the scope of your audit and from there produce a list of patients. Then comes the task of data collection. Make a spreadsheet or database template to work on so data entry is fast and accurate.

Once the data are collected, they need to be analysed and if you have had help from a statistician in designing the audit, they will be able to help you again at this stage. Your seniors will help you to put the results back into a clinical context.

Present your audit formally at the departmental or Trust audit meetings. If your audit findings have been valuable to your Trust, it may be that a wider audience would be interested in them. If this is the case, you may be lucky enough to have the opportunity to develop the study further. This might be in the form of an abstract for a meeting or even a manuscript for publication.

An Example of a Junior Doctor's Audit

There are usually clear Trust guidelines against which the care of patients with urethral catheters on acute medical wards can be audited. A proforma can be devised to ask whether each of these guidelines is indeed being followed. The questions should be kept simple and designed to return binary (yes or no) answers. For example, "Was the catheter identification label stuck in the notes after catheterisation?" This will facilitate the data collection step significantly.

Analysis of the data might identify areas where guidelines, training or even clinical practice can be improved. It may be that placing a urethral catheterisation care plan in every set of notes is the appropriate intervention. These changes can be implemented following your formal presentation, and at an appropriate interval, the practice can be re-audited to assess the effect of these changes. Of course, it may be your successor in your post that completes the audit cycle. However, note that audit is now thought of as a spiral and not a loop since repetition of this process should build higher quality of care at each stage.

Research

What is research?

Research is the process by which a hypothesis is tested. It can take numerous forms ranging from the molecular to the epidemiological, can be quantitative or qualitative, and is almost always hypothesis-based. By the time they qualify, the majority of doctors will have been exposed to the process of research. Whether through basic science lectures, selected study modules, BSc projects, or even MB/PhD programmes, research is an integral part of the study of Medicine at undergraduate and postgraduate levels.

A career in Academic Medicine certainly will not appeal to everyone, however all doctors should have an understanding of the process of research and how its results filter back into patient care in the form of new drugs and new standards. In the same vein, it is important to draw the distinction between audit and research (a frequent interview question). Research tests a hypothesis and its results will inform practice standards, while audit tests practice against such standards.

Why get involved in research?

Aside from the obvious CV benefits, the training in critical thinking, statistical analyses, familiarity with literature searching and scientific writing are all skills that will serve throughout your career irrespective of which route it eventually takes.

The research environment is a great leveler of rank. While a hospital works through the hierarchy of doctors, the research world thrives on ideas, hypotheses and plans to test them. Your idea will be as valuable as the Professor's if you can back it up and present it properly. Most importantly, a close working relationship with senior staff frequently turns into a longer-term mentoring relationship – particularly if you do not fall into the typical junior medical trap of taking on tasks that you do not complete. Clinicians in research have had a lot of bad press largely because the unpredictable and emergency nature of clinical work is such that it always seems to take precedence over the research work. This means that clinicians can be seen to take on too much, over-promising and under-delivering. It is an easy trap to fall into, but an even easier trap to avoid simply by being realistic about your time and your abilities to deliver.

How can you get involved in research?

We have already mentioned it, but you really ought not to think about getting involved in research, audit or teaching until you have your clinical job under control. There is no point having a list of non-clinical accolades if you are not signed off at the end of the post.

It is most unlikely that as a House Officer you will have the time or opportunity to undertake full-scale laboratory research. Unless you have ongoing work from your student projects, or are committing vast sections of your evenings and weekends, the time and concentration required for good quality laboratory-based research is impractical for most junior doctors. Similarly you are unlikely to take the lead and conduct a complete clinical research project largely because of the lead-time involved in setting such a project up. Clinical research frequently requires lengthy ethics applications, careful study design and statistical preparation and new regulations require that people involved in trials undergo Good Clinical Practice training. Despite these hurdles there are still many opportunities to get involved.

So how can you get involved? Think of the process of research and identify the areas where your skill set can serve you most. Your first stop should be one of your seniors – speak to your Consultant, or if you are in or affiliated to an academic unit, a member of staff there.

What are the options?

If you find an opportunity to take on a role in a research project, your contribution will depend on the current stage of the study. You may, for example, become involved in the planning stages, in which case you may be asked to draft a proposal or an ethics application. This will involve a lot of background reading and many revisions, but will engage you with the process and importantly with the principal researchers. The feedback alone will be invaluable. You may wish to take on a more defined task such as designing the study record form onto which data are recorded or even the spreadsheet or database if your skills are more computer-based.

An ongoing project will almost always require the collection of patient samples, and as the only member of staff permanently on the wards, you may be ideally placed to perform this important task. Make sure you get this right. Know when and what samples to take, the transport medium and the arrangements for delivering the samples to the laboratory.

Many junior doctors involve themselves in notes-based research. This case-note review work can be very rewarding, if a little tiring, but make sure you are not simply being used in a secretarial capacity. For example, if your only role is to pull hundreds of case notes then perhaps other members of staff are better placed to do this. If, however, you are asked to review the notes and extract information from them, then while this can be very time-consuming it can be educational and therefore worthwhile. The key is not to be afraid to question the purpose of your allocated tasks and the basis on which they have been designed. For example, "Has a statistical power calculation been performed or has the number of subjects been 'guesstimated'?"

The flux of junior doctors through departments from House Officers through to Senior Registrars means that there are almost always half-completed projects. Perhaps a Registrar had collected data but has gone off to a different job before they had the time to analyse or write up the project. This is the ideal situation to find yourself in. Such work can easily be conducted out of hours, during evenings and weekends and its timely completion is often down to you. Moreover, complete it well and there is likely to be an abstract, presentation or even publication in there for you.

Case Reports

Try to find an interesting case to write up. You may have to take advice from a more senior doctor as to what would be a good case but they are usually very happy for you to produce a draft for them to review. Look for an opportunity where there are lessons to be drawn and always ask yourself whether there is an 'angle' from which you could write. The case does not have to be rare; it can be a common case where there were pitfalls in the diagnosis or management of the patient. For example, in surgery, particularly in big centres, Consultants may want to write up a series of cases where a particular surgical technique was used.

Photographs may be required for case reports or short articles. Get the patient's consent for use of their photograph not only for educational purposes, but also for publication in scientific journals. Often, the journal needs to be mentioned by name on the consent document so list several that you might attempt to submit to. If the journals have their own consent forms for the publication of illustrations you will usually have to gain consent retrospectively.

8. What can House Officers do to progress in their career?

For many of us emerging from medical school into our working careers, getting through the day is as far as our goals go - to begin with. If this continues until the start of the application process for specialty training, big decisions about career paths really will need to be taken rather abruptly. Some people seem to know from the outset what specialty they want to enter into. However if you, like many, have not started with a particular career within medicine in mind, you will soon realise that you will have to make some important choices from a relatively early stage. The next few pages will outline some of the thought processes that can help you decide the direction in which your career can go and where to get help when you need it.

The Training Structure

Medical training has undergone some radical changes over the past few years with many competing demands on junior doctors' training. While these changes have altered the nomenclature and even the timing of training, the principles of selection into basic and then higher specialist training remain the same.

Specialty Training follows on from the Foundation Programme and applications for these schemes are where all the panicking begins! It is true to say that compared to older systems, junior doctors are required to commit to training within a given specialty at an earlier stage.

Entry into Specialty Training involves an application process which constitutes an application form and interview. In some specialities there will also be an exam. Whilst some specialties have maintained a 'run-through' structure from entry following Foundation Year 2 (for the time being, at least), many specialties now have a Core Training base and require open competition by way of further application forms and interviews to proceed to ST3 and beyond. On completion of this programme, trainees are awarded the Certificate of Completion of Training (CCT) and are eligible to enter the Specialist Register or the General Practitioner Register.

Your Career

Many young trainees do not know what to expect from their careers, but remember that it is *your* career. This means that there is not a right or a wrong way to go about it. Discussions about careers do not happen nearly often nor openly enough in medicine. We think we are too busy to think about anything other than the next four months or that our seniors are too busy or too important to give us the time for advice. Both of these assumptions are simply wrong. Just like in any other career, you need to know where you want to go and then you can decide how to get there.

Do seek advice and guidance but in the end your reasons for choosing a particular discipline or specialty will be very personal. Some of the most successful people in their fields stumbled into those careers for the most prosaic and seemingly insignificant reasons.

The Application Form

Progression through the different stages of your career will be almost always be governed by the ubiquitous job application form. This selection tool aims to identify strong candidates by asking information that one might usually find on a curriculum vitae, namely:

Personal Details
Medical School – Degrees, Distinctions, Prizes
Postgraduate Exams
Research Experience
Audit Experience
Teaching Experience
Presentations at Meetings
Publications

The form will also assess a candidate's suitability and commitment to their chosen career through a series of short answer open questions such as:

Give details of your clinical experience and level of competence.
Describe an audit in which you were most involved and how it was effective.
Give a recent example that demonstrates you possess communication and interpersonal skills.

Devising a strategy to tackling these forms is beyond the remit of this chapter; however it is always worth remembering that the shortlisting and interview panels are not primarily concerned with the facts in your answer. The assessors are not marking the answer to your 'ethical dilemma' question on how good story is, for example. Rather, they want to know how well you can analyse a situation and apply an ethical framework. They are not marking how dramatic your 'breaking bad news' scenario was. They are much more interested in whether you were able to conduct yourself appropriately and empathetically. In a 'clinical error' question the magnitude of the error is not being assessed, it is your understanding of your responsibilities and of your probity that will determine your mark.

As such, you will be marked on how you describe your experience and what you have learned from it not solely the content. The key is in your style of writing. Some doctors moan about this, saying that how well you write cannot be a marker of how good a doctor you are. This attitude is unhelpful on two fronts. Firstly, that is the system, so just get on with negotiating it to the best of your ability. Secondly, the ability to present (any) information succinctly and effectively is the cornerstone of good clinical practice.

Knowing that there is an assessment of your portfolio, that there is a form to fill in and an interview ahead is a great way to focus the mind towards ensuring that appraisals go smoothly and none of the boxes on the application forms remain embarrassingly blank. Publications and prizes count as much as practical experience and procedural skills.

Start now. There is no point worrying about not having any prizes when applications open – you may only have a couple of weeks to submit your form! Start thinking about how you can build your profile generically, irrespective of whether you do or do not know where you want to end up.

Knowledge and skill

As unfashionable as it has become to say, you really ought to be learning, academically as well as practically, from every patient you meet. It is possible to overlook the importance of acquiring the knowledge-base and skill-set you need to progress through a medical career. To be truly valuable to your patients and therefore have a successful career, you should aim to acquire the highest standards of knowledge and skill. It is a sad fact that Medical Schools have consciously removed large chunks of basic and clinical science from curricula and many Foundation doctors doubt their own knowledge base as a result. You should take confidence in the fact that you will not have been allowed to graduate had you not known enough to practice as a House Officer, but make every effort to identify these 'blind spots' and rectify them.

Learn from every patient

See as many patients as possible on the ward round, in pre-assessment clinic or on an acute take. Make it a habit to reflect on your clerking and management plans. Ask yourself, "Is there anything I could have done better?" Ask for feedback from your peers - what things would they do differently and why? If you come across a case where your knowledge is lacking or that you just find interesting, read up on it.

Quickly jotting down an incident in a notebook or diary will serve you in terms of day-to-day learning and also will remind you of interesting scenarios around which you can answer application form and interview questions.

Seek out opportunities to present your cases at Morning Reports, departmental meetings or Grand Rounds. You may even be able to write certain cases up as case reports – an increasingly valued currency when it comes to application forms. The feedback you receive from a senior colleague when you are about to present a case in public with his or her name on the first slide can be surprisingly more thorough and insightful than when the same colleague is asked to complete a 'Case-based Discussion' assessment.

Practical skills

Try to grab every opportunity to do procedures as the more you are able to do, the more confident you become and the less you rely on your Registrar! Enjoy the fact that as a House Officer you are *expected* to be inexperienced at more complex procedures, so you can only benefit from putting yourself forward for doing them.

Sometimes it may be difficult to get many opportunities in a teaching hospital but do not despair. There are always procedures to be done. A good tip is to let Registrars know what procedures you would like to do and leave your bleep number with them. Most of the time they will keep you in mind and gladly guide you through the procedures when opportunities arise. If somebody offers you the chance to do a procedure, never turn it down. You do not know how long it will take for the next opportunity to arise, and although this may mean staying late after work, or even later after an on-call, it will be worth it not only for the practice that you will gain but for the fact that you will have demonstrated to you seniors that you are committed.

Teaching

This is great way of keeping your knowledge and skills up-to-date. There are always medical students lurking around the wards desperate for someone to teach them – you were that student a few short months ago!

Provide informal bedside teaching in the evenings after work. This will not only encourage you to do some reading but will refresh your clinical skills too. Ask for some formal feedback which you can use to both improve your teaching skills and add to your portfolio as proof of your teaching. You could produce a short feedback form yourself which you can ask the students to fill in at the end of the session. (You are unlikely to get formal feedback unless you set the ball rolling yourself!)

Exams

Advice here can be conflicting. Some trainers seem to encourage Foundation doctors not to sit membership exams early, perhaps favouring workplace-based assessment. However, membership of British Royal Colleges remains an internationally recognised standard of knowledge and skill. Bearing in mind that assessment drives learning, if you know which discipline you wish to pursue you should get on and study for the exams. When you feel confident in your knowledge, then you should enter for them. Properly prepared, there is no reason why you should not pass and having your Part I behind you is a great way to show your commitment to your chosen discipline when it comes to applying for ST1 or CT1. Moreover, given that you should aim to complete the membership examinations in advance of ST3 application, you really do need to start as early as you can.

Research & audit

You should seek out opportunities for conducting investigative work, be that research or audit. Chapter 7 covers this area fully.

Clinics and procedure lists

There is a misconception that clinics, procedure lists and theatre are off limits to House Officers. This, again, simply is not true. Your first responsibility is of course to your ward patients.

However not all jobs are super-busy and if they are, certainly not all of the time. It is worth asking to attend clinics and lists. For most specialities the care of in-patients forms only a small part of the workload and you should know what else specialists do if you are to make your own informed career choices and demonstrate why you have made them.

You may feel that these things are impossible to organise but they are actually easy to sort out. All it needs is the confidence to ask a friendly Consultant whether they would be happy for you to attend. Most of the time, Consultants and Registrars will be impressed by your drive and would be only too happy to get you along to their other activities. Taking the initiative to attend clinics and procedure lists certainly shows commitment to a specialty. Even if you are not allocated a Foundation rotation which includes the specialty you believe you would like to train in, you can still show commitment by arranging to spend your "Taster Week" there in your second Foundation year. Another way of showing commitment to a specialty is by attending relevant meetings, courses or conferences. Given that the study leave time Foundation doctors are entitled to is taken up in many Trusts with formally allocated teaching time, it may be worth sacrificing one or two days' annual leave in order to do this.

Mentors

A mentor is usually someone who inspires you and who takes an interest in you and your plans. They can help by giving you advice, motivating and perhaps pointing you in a direction that you would not have thought about if left alone. Importantly, they will be the people you really want to keep in touch with after you have left a job, and they will want to hear from you. All your seniors will have been in the same position as you and therefore are more than happy to share their successes with you. A good mentor will be particularly useful in putting you in touch with people other than themselves who can offer you learning opportunities, but the impetus has to come from you. Ask for what you want.

Making the choice

While certain jobs, certain colleagues and certain encounters will help you learn what you do (and importantly what you do not) want to do, only you will be able to make the important choices. Rather than letting this feel onerous, you should relish the opportunities you have over the first 18 months of your Foundation Programme to make your choices. You may have thought when you started Medical School that becoming a doctor was the biggest career choice you had to make. However, it is only now that you can properly appreciate all the options that lie before you. Whether it is primary care, hospital medicine, public health, academia, service provision or health policy that fascinates you – you have the luxury of choice.

Start thinking about your career early, set a personal goal and always ask for advice. More importantly, whatever career path you choose, make sure it is something you absolutely love and would not mind waking up in the morning (and sometimes the middle of the night) for!